QL49 .S596
Simon, Seymour. Animals in your neighbor

Y0-CSW-175

SIMON **CHILDREN'S LIT.**

Animals in your neighborhood

DATE DUE			
DEC 2			
APR 30			
MAY 22			

Animals in Your Neighborhood

Seymour Simon

*illustrated by
Susan Bonners*

**WALKER AND COMPANY
NEW YORK**

Text Copyright © 1976 by Seymour Simon   Illustrations Copyright © 1976 by Susan Bonners
All rights reserved. No part of this book may be reproduced or transmitted in any form or by any means, electronic or m
anical, including photocopying, recording, or by any information storage and retrieval system, without permission in wr
from the Publisher.
  First published in the United States of America in 1976 by the Walker Publishing Company, Inc.
  Published simultaneously in Canada by Fitzhenry & Whiteside, Limited, Toronto.

Trade ISBN: 0-8027-6270-0   Reinf. ISBN: 0-8027-6271-9   Library of Congress Catalog Card Number: 76-11904
Printed in the United States of America.   10 9 8 7 6 5 4 3 2 1

Observing Animals in Your Neighborhood

Walk around your neighborhood. You may be surprised at all the animals you see even in a big city. Besides dogs and cats, you may see squirrels, flocks of pigeons and starlings, sparrows, insects, spiders, and lots of clues to other animals.

You will have fun and find out more if you take the time to record the observations you make. You may remember what you see after a few days, but what about after a few weeks, or a few months? Jot down notes on cards or in an observation log. Record the time of day, the place, the weather, and exactly what you see.

Drawings and photographs will often help you remember even more accurately than notes alone. You do not have to be an artist. Just sketch some of the important points so that your notes will be a better reminder. If you take photos, try to get as close to the object as you can.

Look carefully at tracks, nests, and other clues to animals that you find. Collect a feather for later identification. Learn to make casts of animal tracks in mud.

Above all else, be curious about what you see around you. You don't have to travel to distant places to be an explorer. You can observe and find out about animals close to home. Your own neighborhood is a part of the world of nature.

One day go for an animal hunt in your neighborhood.
See how many different kinds of animals you can spot.
Dogs, cats, and squirrels are common animals
in many neighborhoods.
But not all animals are four footed and furry.
Birds, frogs, and snakes are animals.
Ants, spiders, butterflies, and earthworms also are animals.
Every living thing that is not a plant is an animal.

On your animal hunt, look on the ground, on bushes, on roofs and window ledges, and in trees.
Look all around you.
Look for animals in the morning, in the afternoon, and at night.
Look for them when it rains and when the sun is shining.
Listen, too, for sounds that animals make.

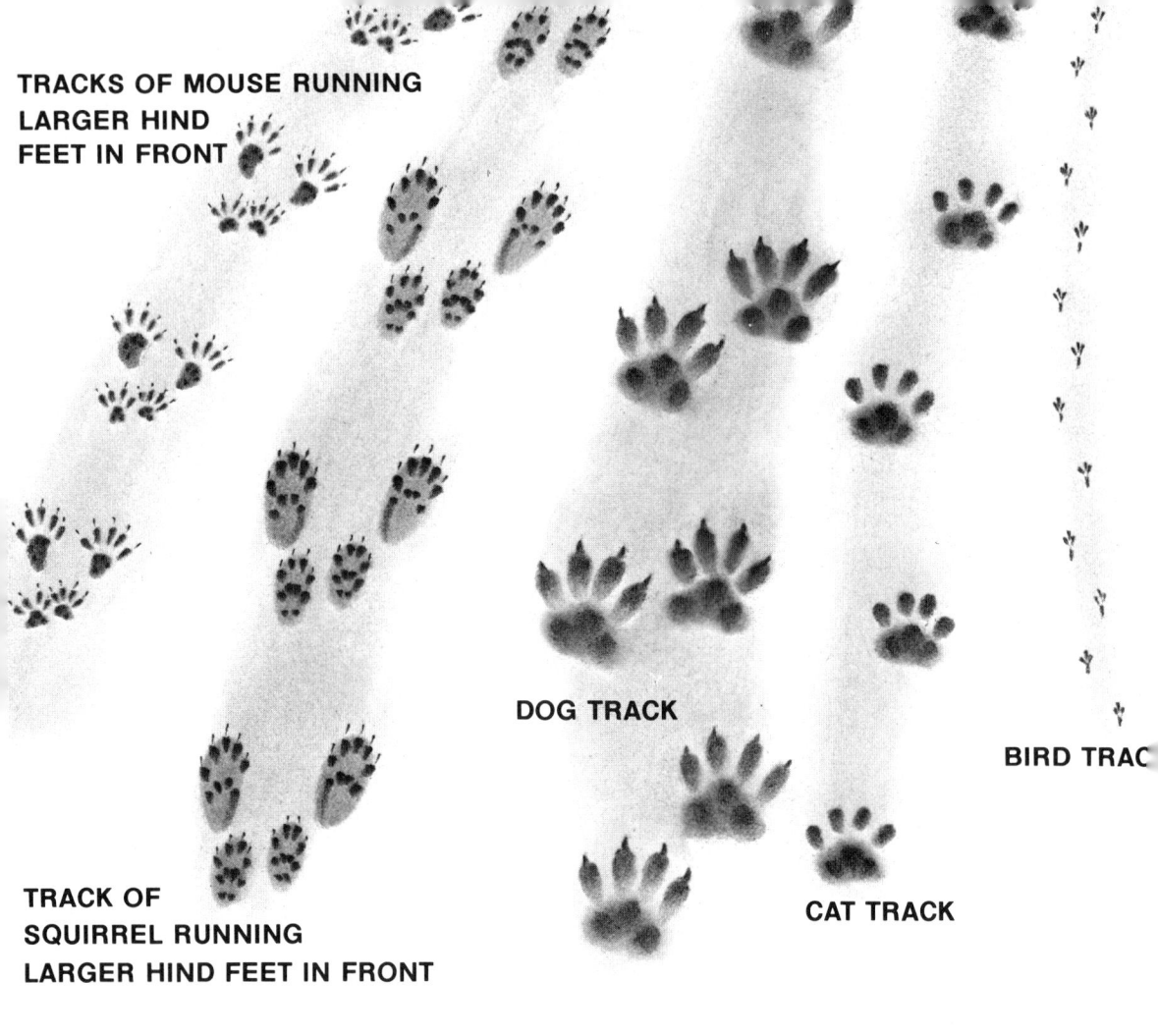

**TRACKS OF MOUSE RUNNING
LARGER HIND
FEET IN FRONT**

DOG TRACK

BIRD TRACK

**TRACK OF
SQUIRREL RUNNING
LARGER HIND FEET IN FRONT**

CAT TRACK

Sometimes animals leave signs or clues that they
have been in your neighborhood.
A feather shows that a bird has been visiting.
A pile of empty nut shells shows that a squirrel was nearby.
A web shows that a spider was there.
Which animals might leave clues such as these:
a cocoon, a hole in a leaf, a mound of soil?
Look also for animal tracks in snow or mud.
Sometimes you can even see tracks that were left
in sidewalk concrete.
What other clues to animals can you find?

Is there a park or a vacant lot in your neighborhood?
Pick a small grassy patch to observe closely.
Get on your hands and knees and look in the grass,
under a pile of leaves, and under a rock.
Do you see any insects?
An insect is a small animal with three parts to its body.
It has three pairs of legs and
two feelers called *antennae* on its head.

MOTH

Most insects have wings but sometimes they are folded down and hard to spot.
Beetles, flies, mosquitoes, cockroaches, grasshoppers, crickets, butterflies, and moths are insects.
How many wings does a fly have?
How many wings does a butterfly have?
Check on the number of wings of any insect you find. It helps to tell them apart.

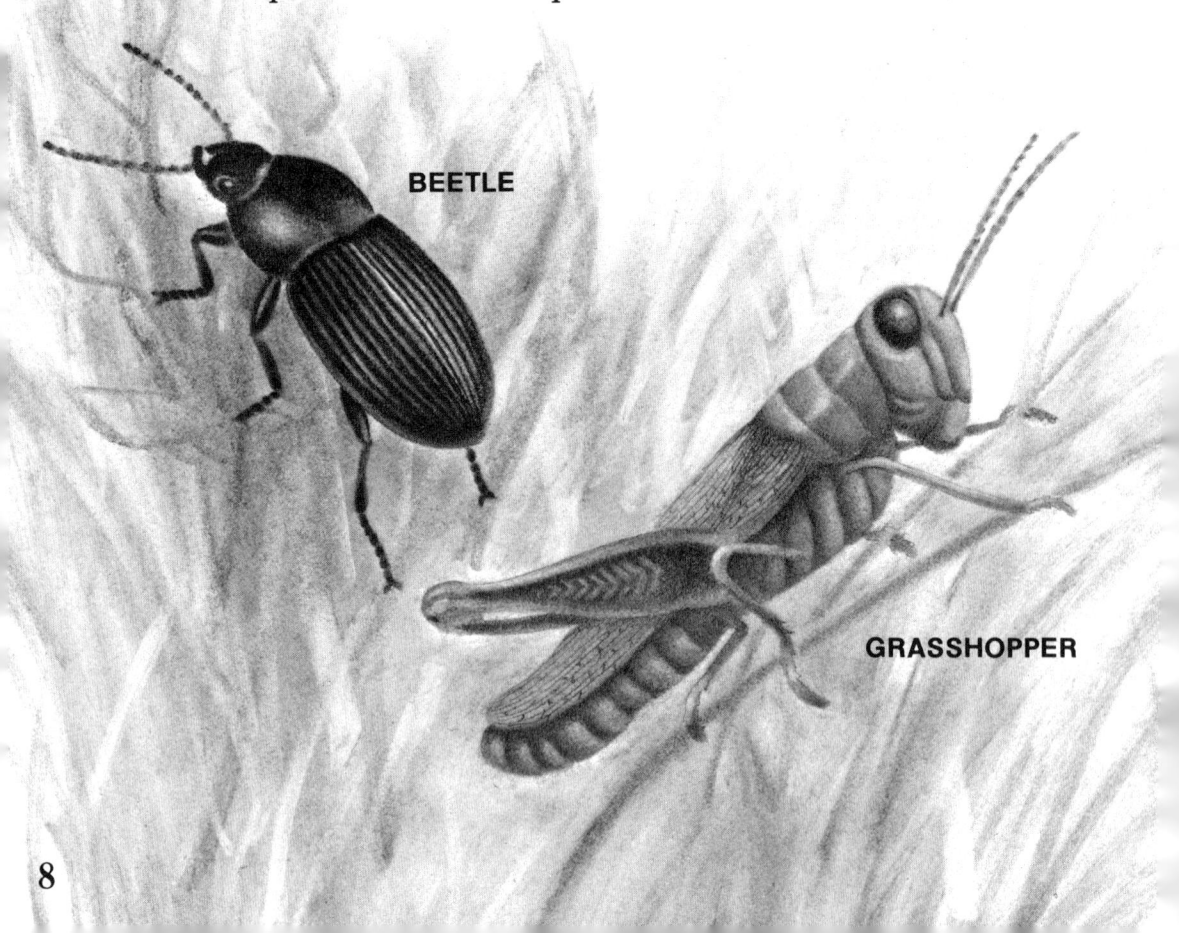

BEETLE

GRASSHOPPER

CABBAGE BUTTERFLY

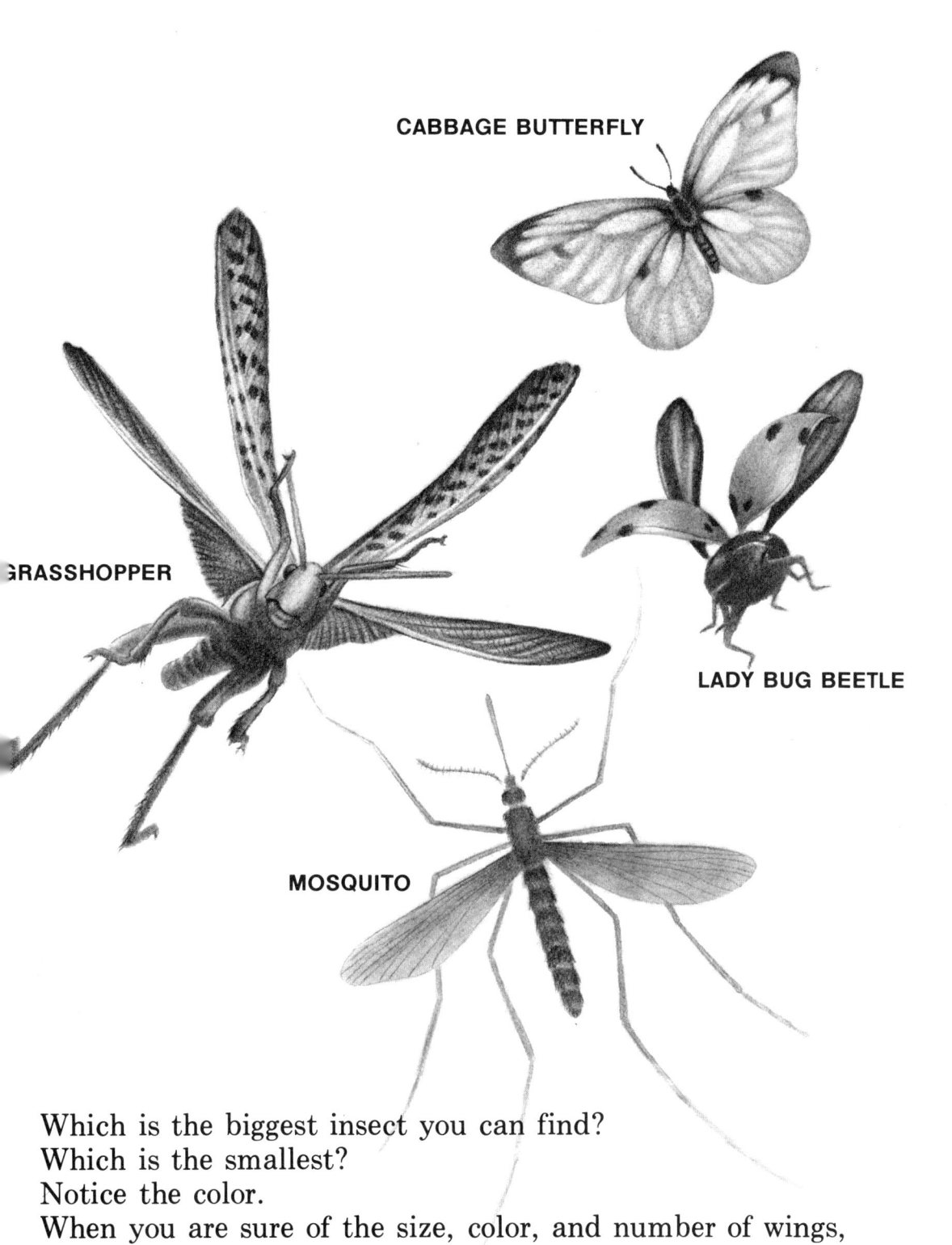

GRASSHOPPER

LADY BUG BEETLE

MOSQUITO

Which is the biggest insect you can find?
Which is the smallest?
Notice the color.
When you are sure of the size, color, and number of wings, look up the insect in a guide book.

9

Notice how insects move.
Do they crawl or walk or fly?
See if you can find out what they eat.
Can you hear any insect sounds?
At night you may hear crickets, cicadas, and katydids.
Katydids make sounds just like their names.
Crickets chirp at a regular rate.
Cicadas make a steady hum.
Follow the sound till you feel you are very close,
then use a flashlight to try to find the insect.

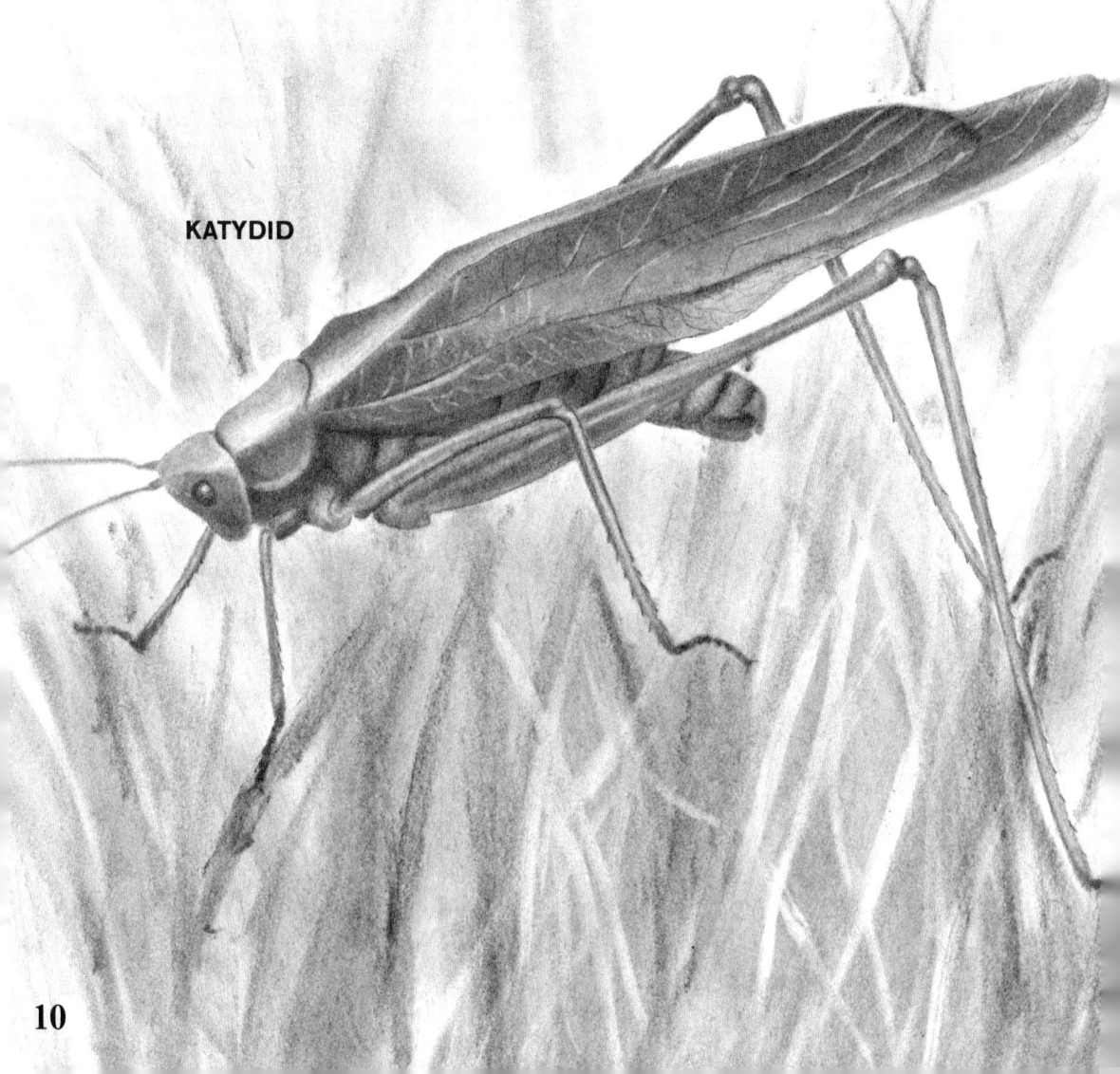

KATYDID

CICADA

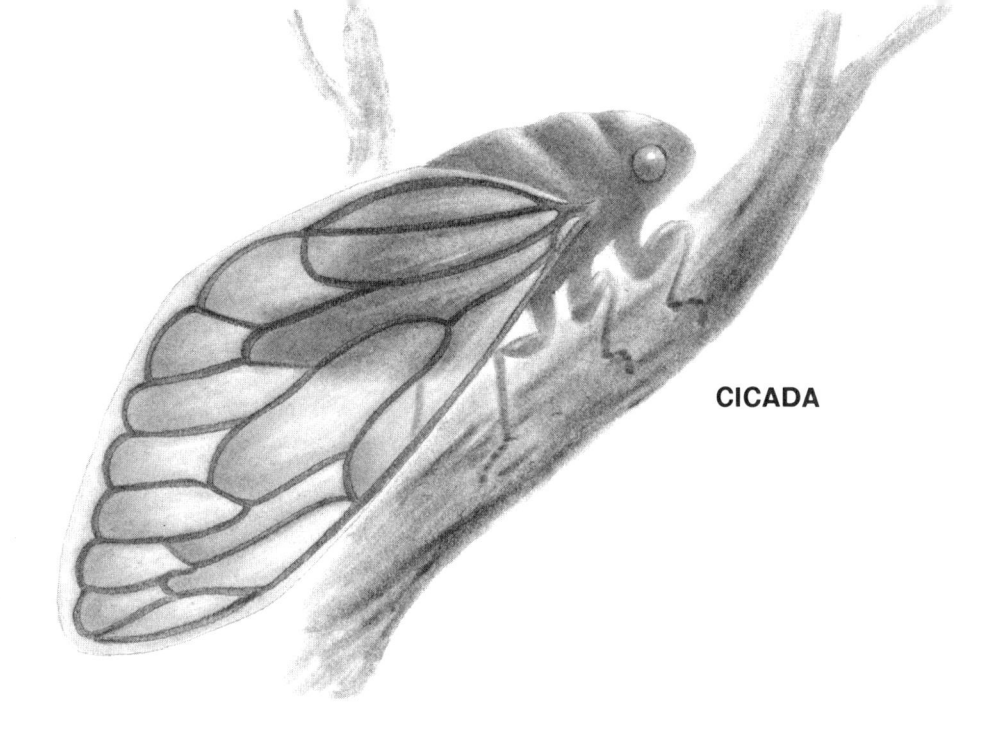

Look for insect eggs in the grass and in the soil, in the bark of trees and on leaves.
What other clues to insects do you see?

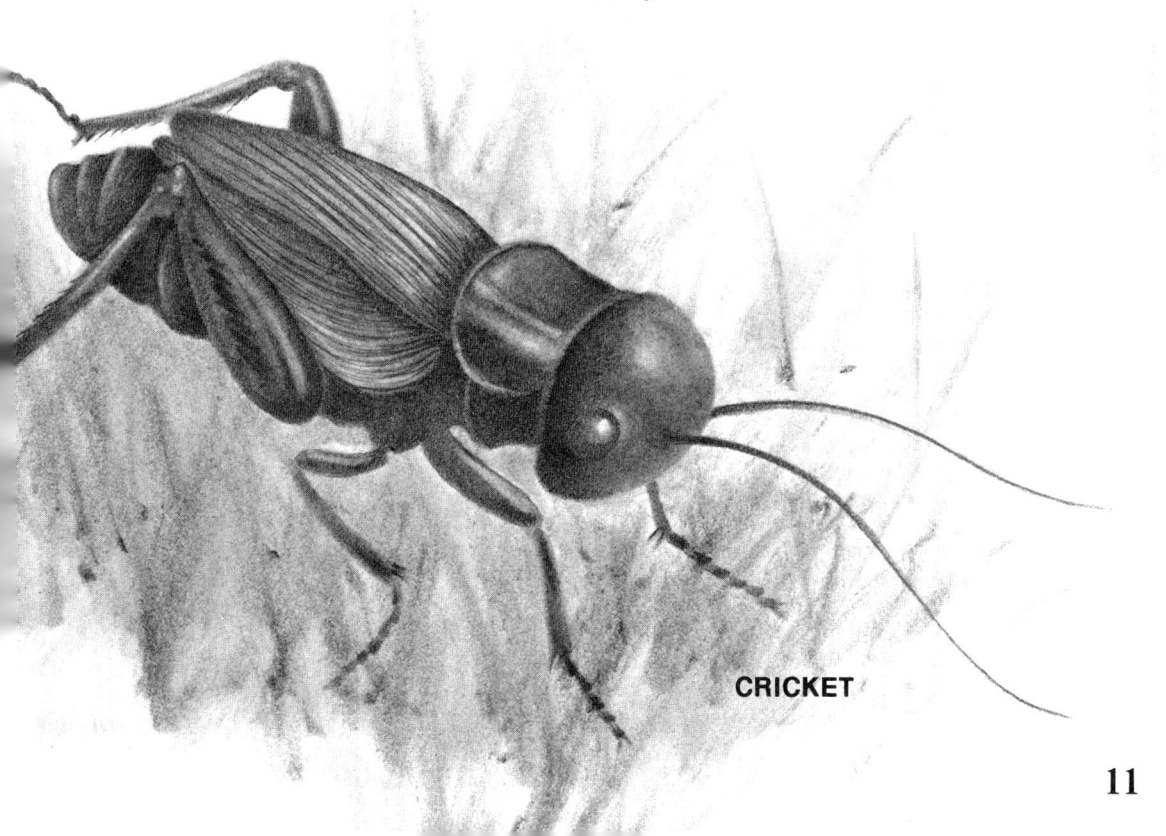

CRICKET

**HOUSEFLY
LANDING ON CEILING**

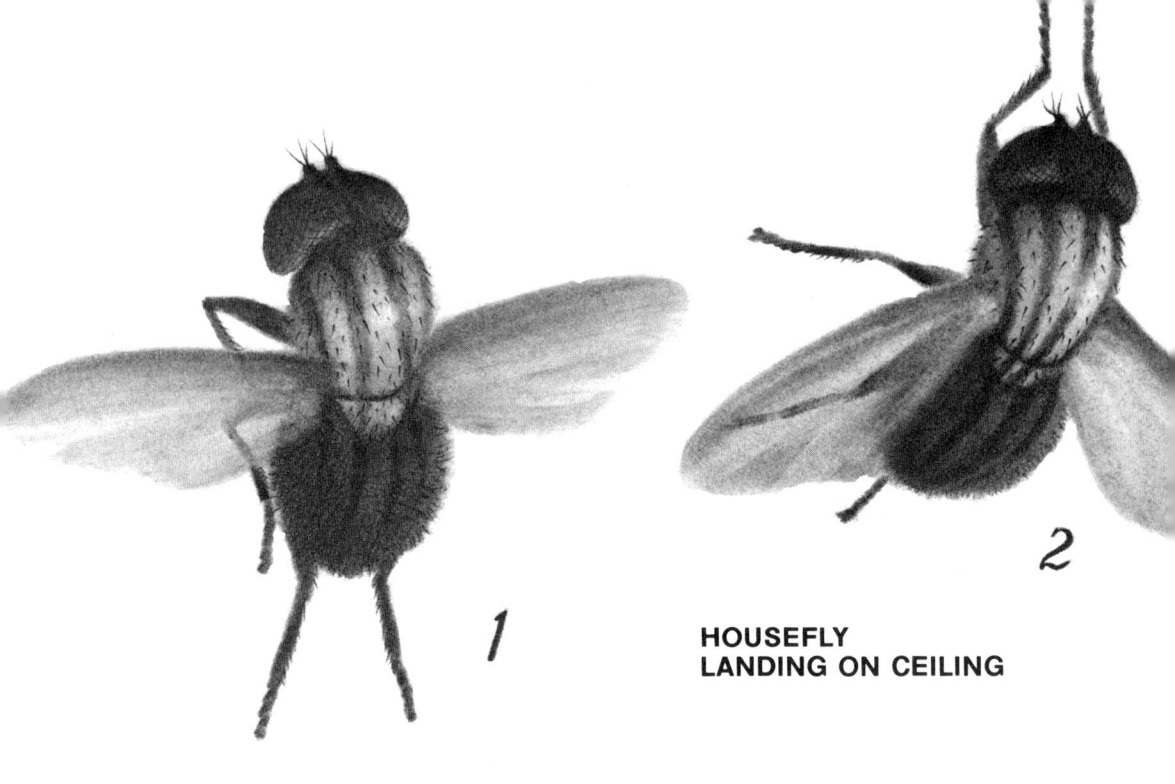

Flies are insects found almost everywhere
in cities, suburbs, and out in the country.
One kind of fly is the housefly.
Look at a housefly closely.
Use a magnifying lens if you can.
The hair on a housefly's legs catches onto droppings
and bits of food.
In this way it carries germs from one place to another.
Sometimes flies land on food that is not covered.
Germs may get on the food and make you sick if you eat it.
Watch how a fly is able to walk up any smooth surface
such as a pane of glass.

3

A fly has little claws and sticky pads at the end
of each of its six legs.
These hold on to almost any surface.
Its two large eyes are made up of many little eyes
close together.
A fly will react quickly to any nearby motion.
With its small wings it can take off instantly,
fly in any direction at great speed,
and land in any position, even upside down.
It is no wonder the fly is so successful
at living in many different neighborhoods.

Another insect that you probably know is the cockroach.
All roaches have wings, but they rarely fly.
To escape their enemies, roaches depend
upon their ground speed.
They are very fast for such small animals.
Cockroaches live in dark places wherever they can find water.
They can eat almost anything, from leftover bits of food
to the glue in the spines of books.
During hard times, roaches will eat other roaches.
Even without any food, a roach may live for weeks or months.

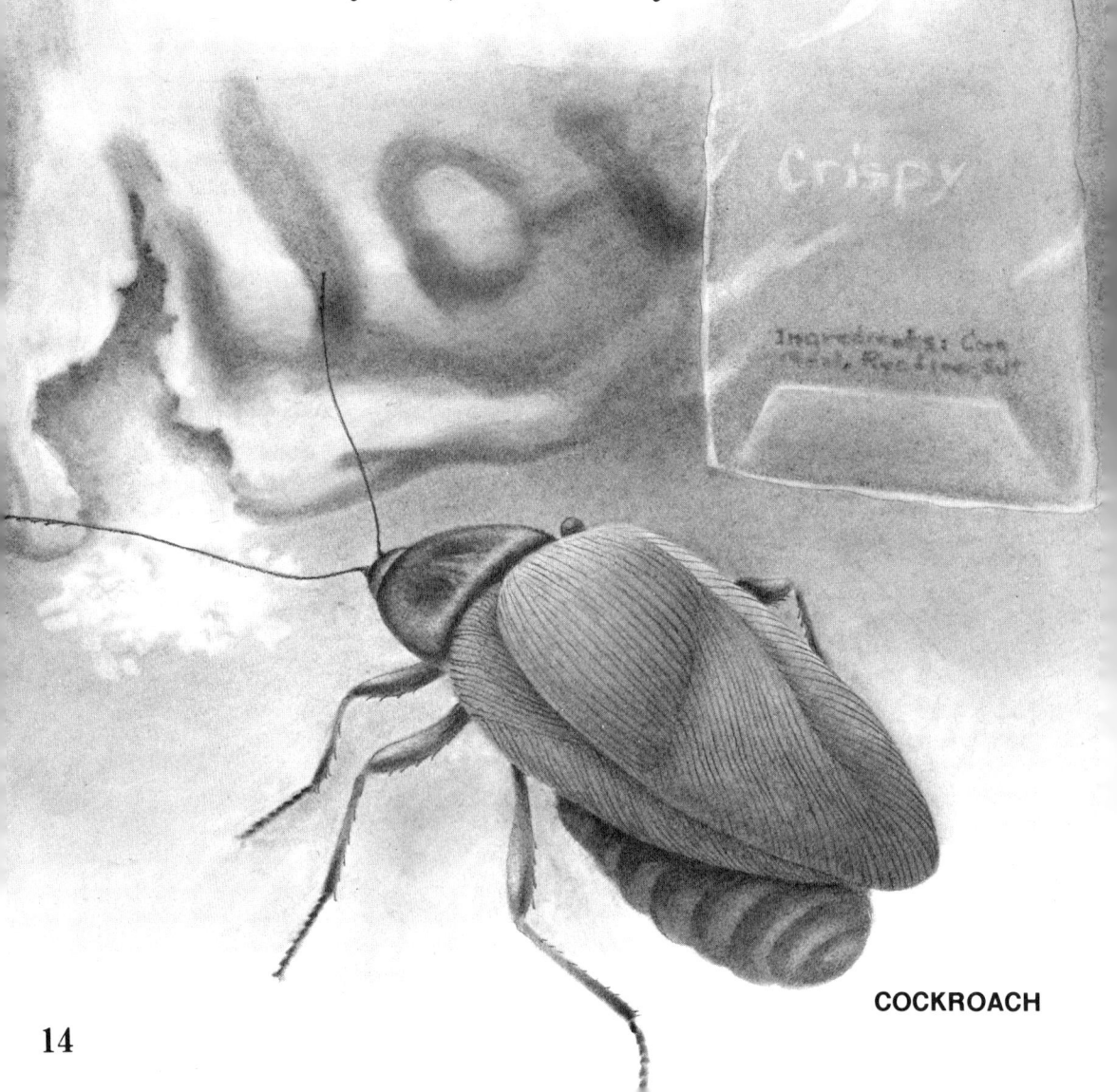

COCKROACH

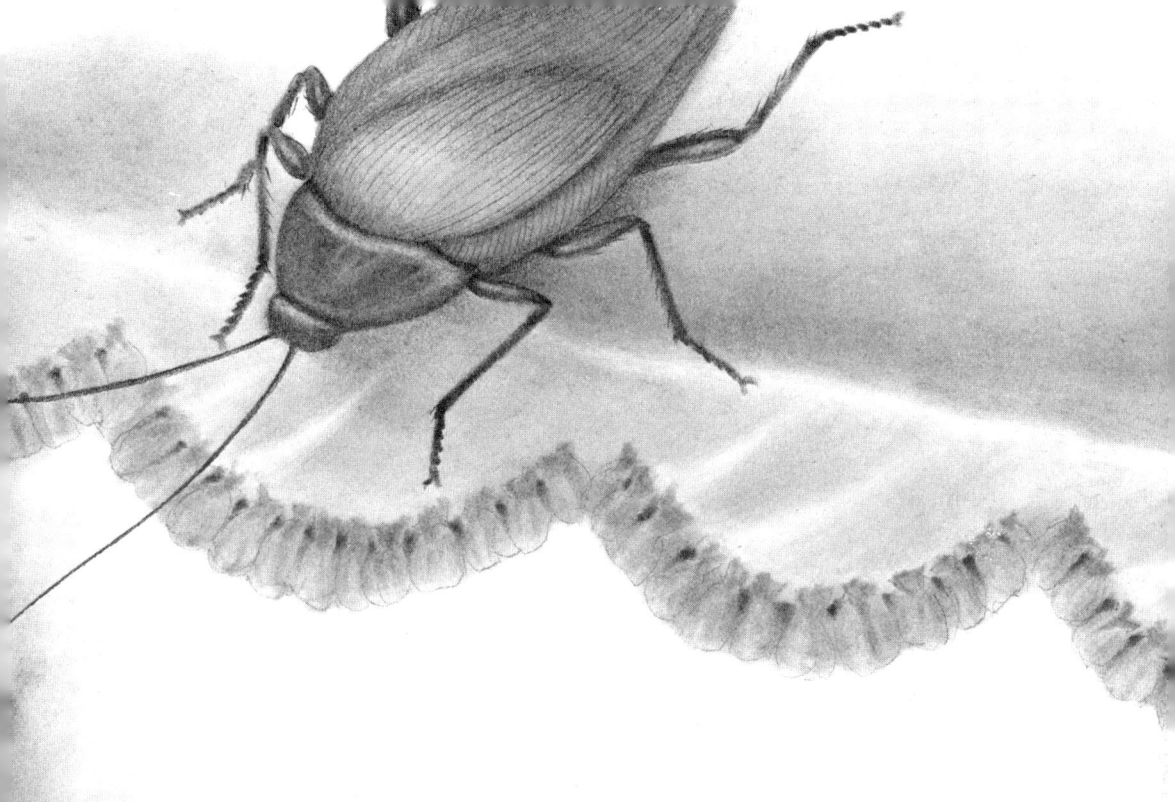

Under good conditions of plenty of food and water, roaches multiply quickly.
A female roach can lay eggs about seven times during her life. Each time about three dozen young may emerge from the egg case. It is no wonder that cockroaches have survived on earth for many millions of years.

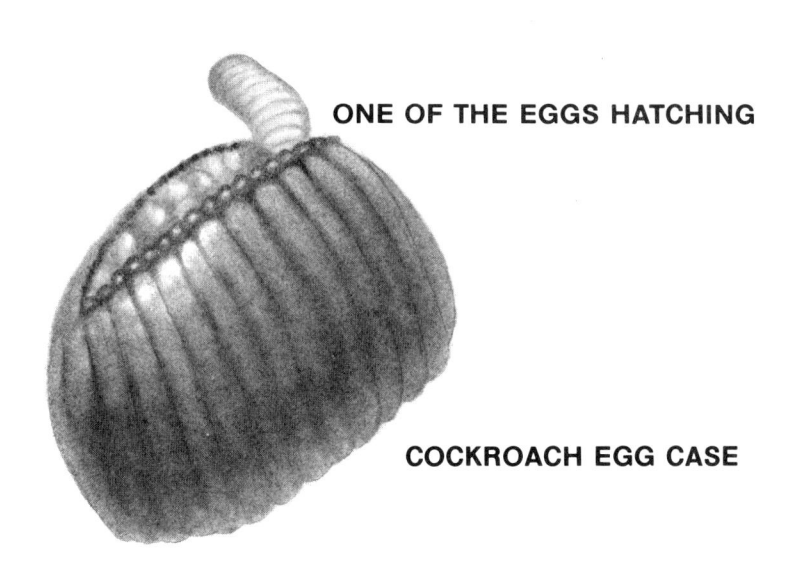

ONE OF THE EGGS HATCHING

COCKROACH EGG CASE

One insect that you are almost sure to see is an ant.
Perhaps you will see more than one kind.
How are the ants you find different from each other?
How are they the same?
Watch an ant carrying something.
How big is the ant?
How big is its load?
Can an ant carry something as big as itself?
Can you carry something as big as yourself?
Do you think an ant is strong for its size?

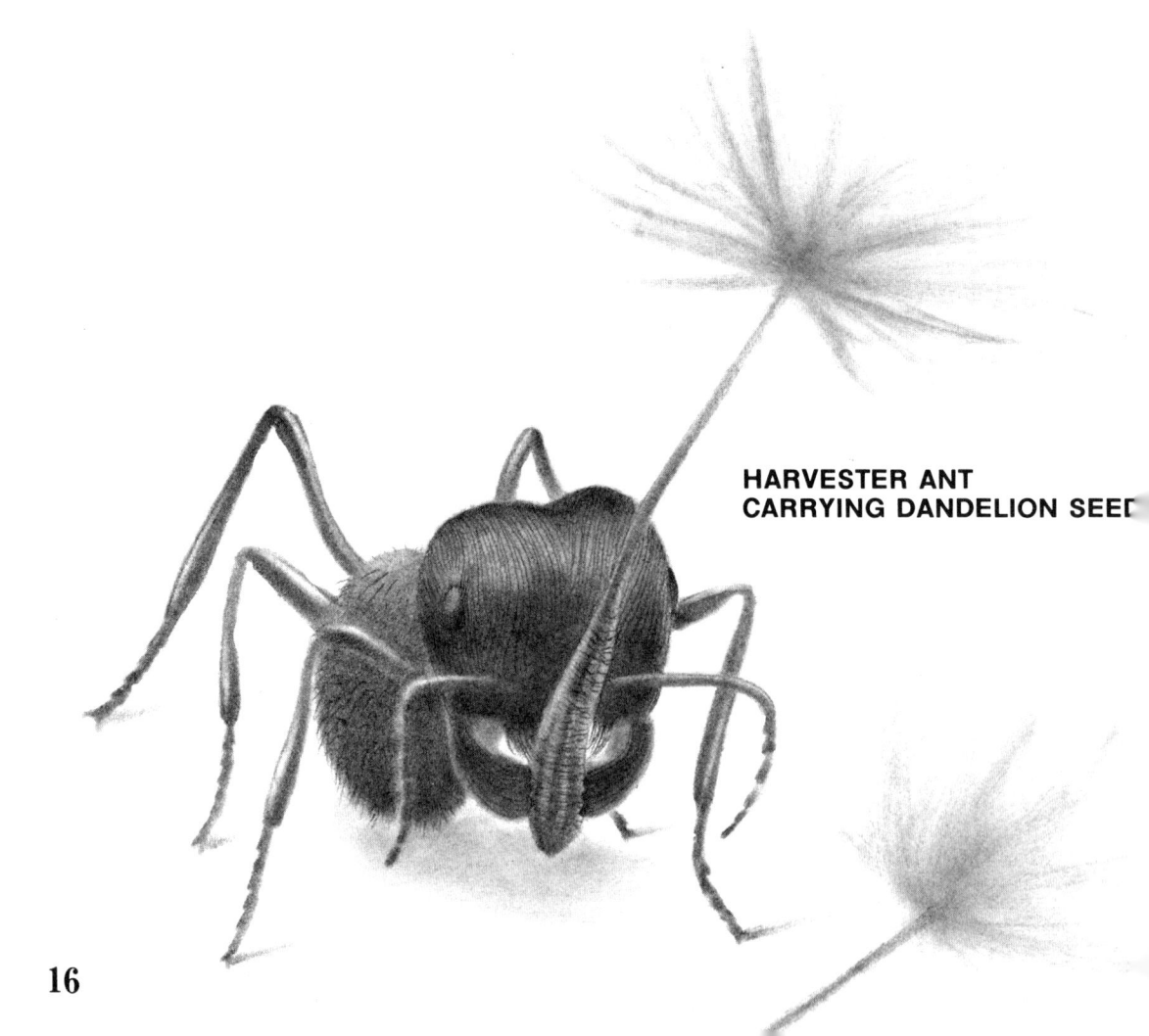

**HARVESTER ANT
CARRYING DANDELION SEED**

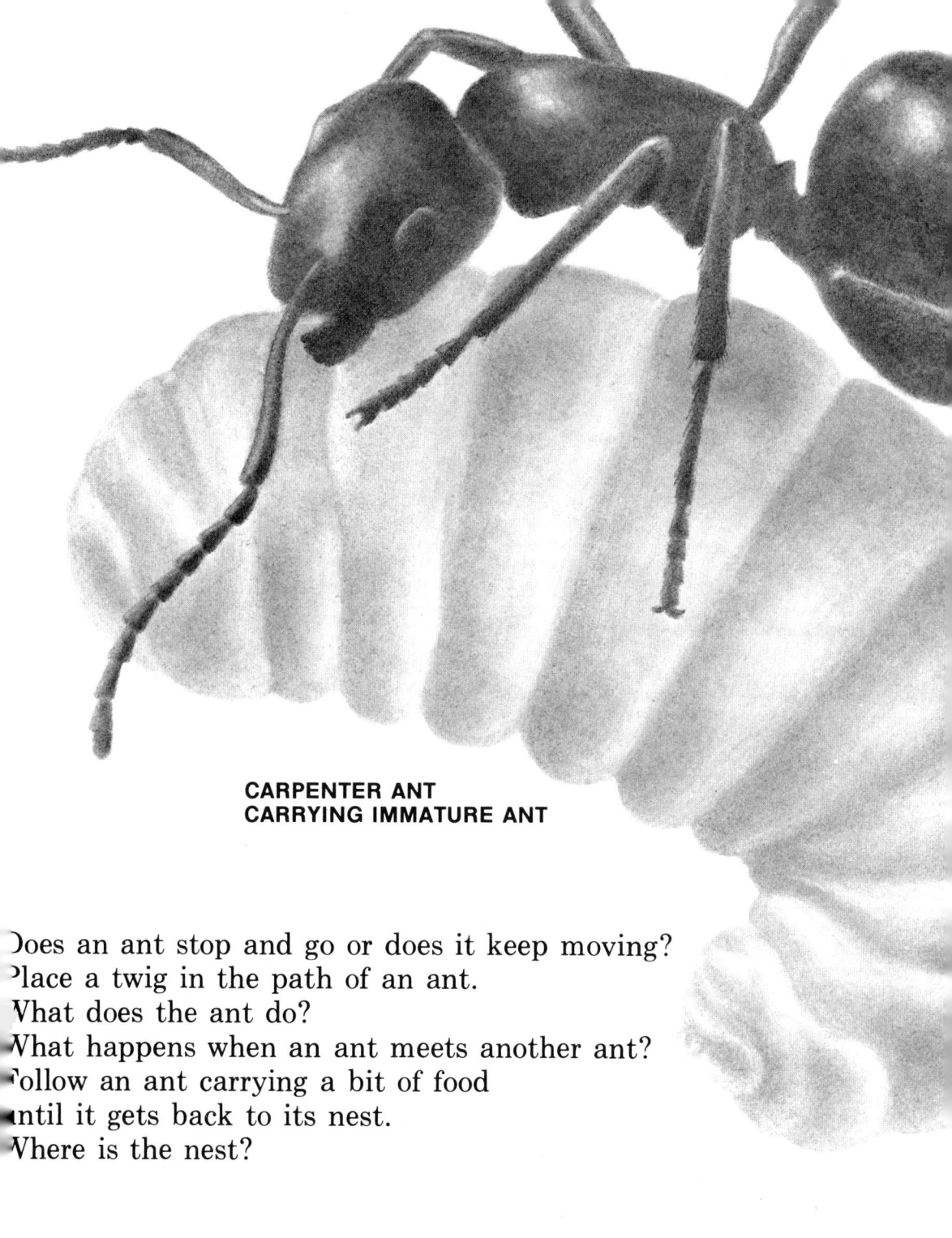

**CARPENTER ANT
CARRYING IMMATURE ANT**

Does an ant stop and go or does it keep moving?
Place a twig in the path of an ant.
What does the ant do?
What happens when an ant meets another ant?
Follow an ant carrying a bit of food
until it gets back to its nest.
Where is the nest?

Some ants live in mounds on top of the soil.
Other ants live in underground nests.
Still other ants live in the trunks of trees
or in the wood beams of houses.
Place a bit of food a short distance from an ant nest.

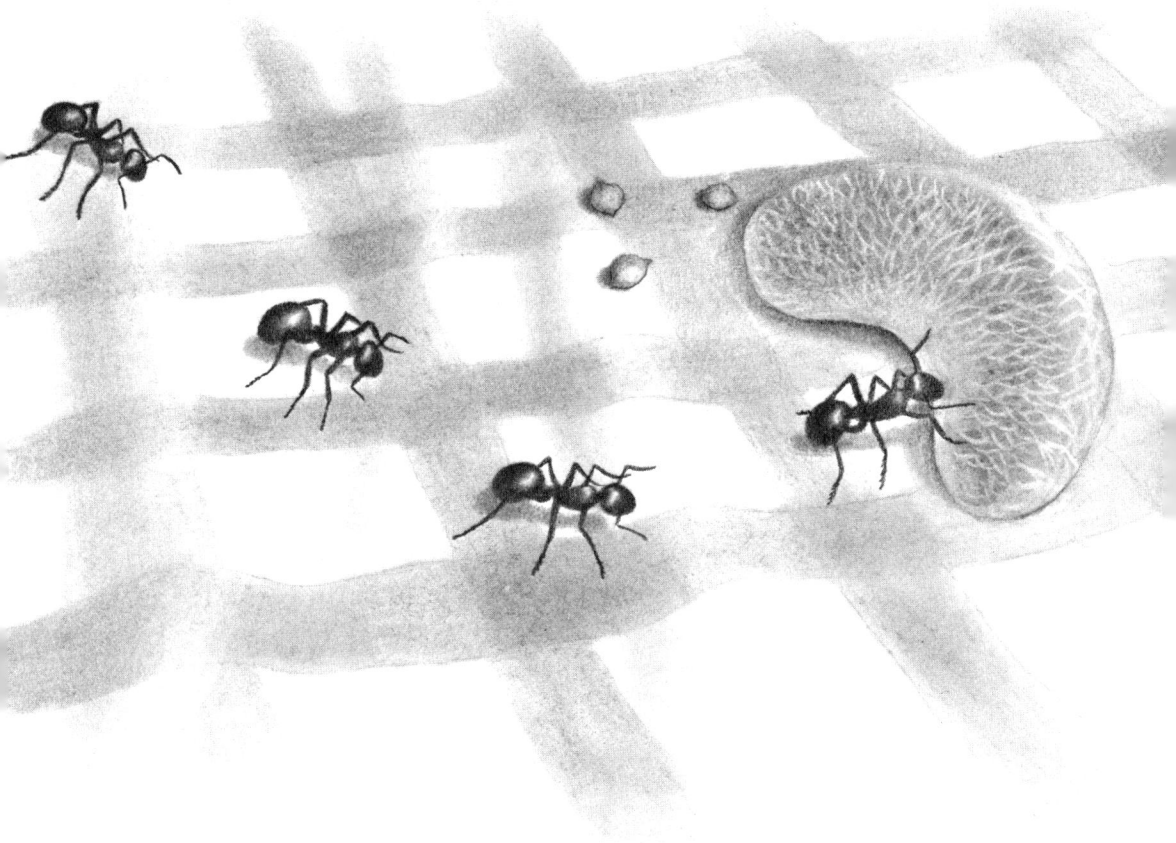

Watch to see how long it takes
for the first ant from the nest to find the food.
What does the ant do?
Watch closely to see it touch the end of its abdomen
to the ground.
It is depositing an odor trail that the other ants can follow.
Do other ants come soon?

**CARPENTER ANTS
TOUCHING ANTENNAE**

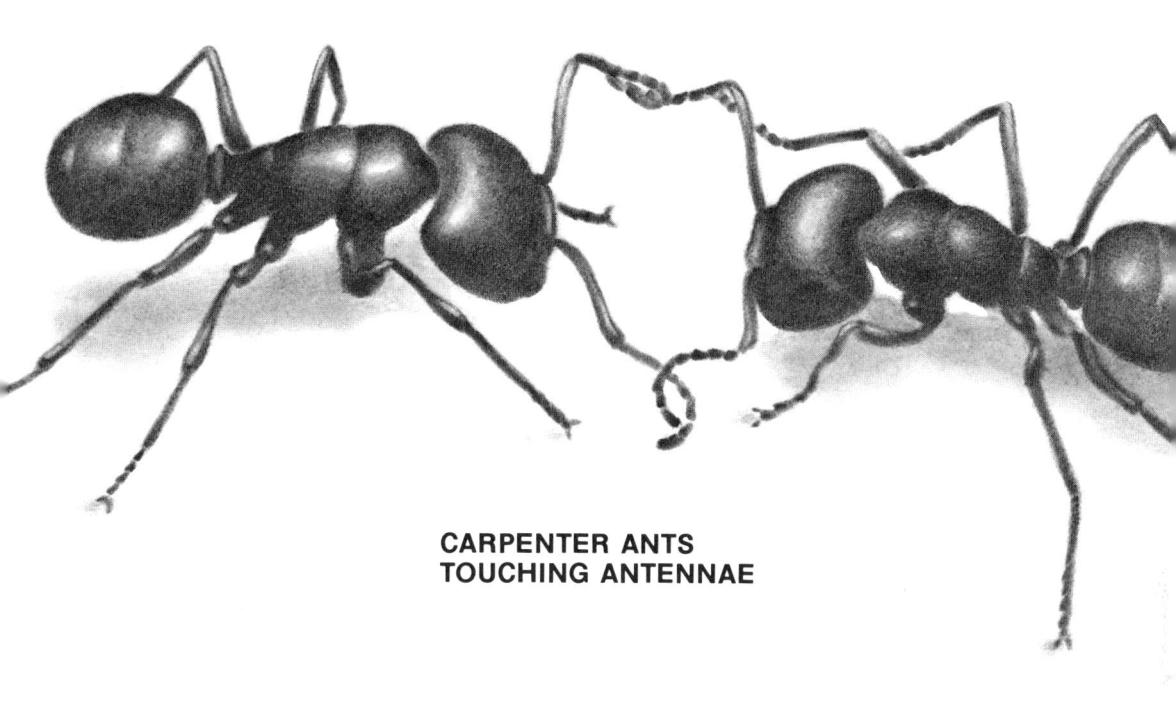

What happens when several ants
come upon the food at the same time?
Place bits of different kinds of food near a nest
to see which kinds ants eat.
Try sugar, a bit of egg or meat, and a tiny piece of lettuce.
When winter comes, ants go deep within their nests.
When do you see the last ants of the year in your neighborhood?
When do you see ants again in the spring?

**CARPENTER ANT
JUST BEFORE
ADULT STAGE**

Not all the tiny animals that you see are insects.
Unlike insects, spiders have eight legs and two body parts.
You may see a spider in your neighborhood
climbing up a wall or scurrying through the grass.
Some spiders spin webs to catch their prey.
You might even find a web in a corner of a room
in your house.

GARDEN SPIDER

WOLF SPIDER

BANDED GARDEN SPIDER

Other spiders hunt for their prey along the ground.
If you find a web, touch it lightly with a twig.
Perhaps the spider will come out to see
what caused the disturbance.
Can you find more than one kind of spider in your neighborhood?

CENTIPEDE

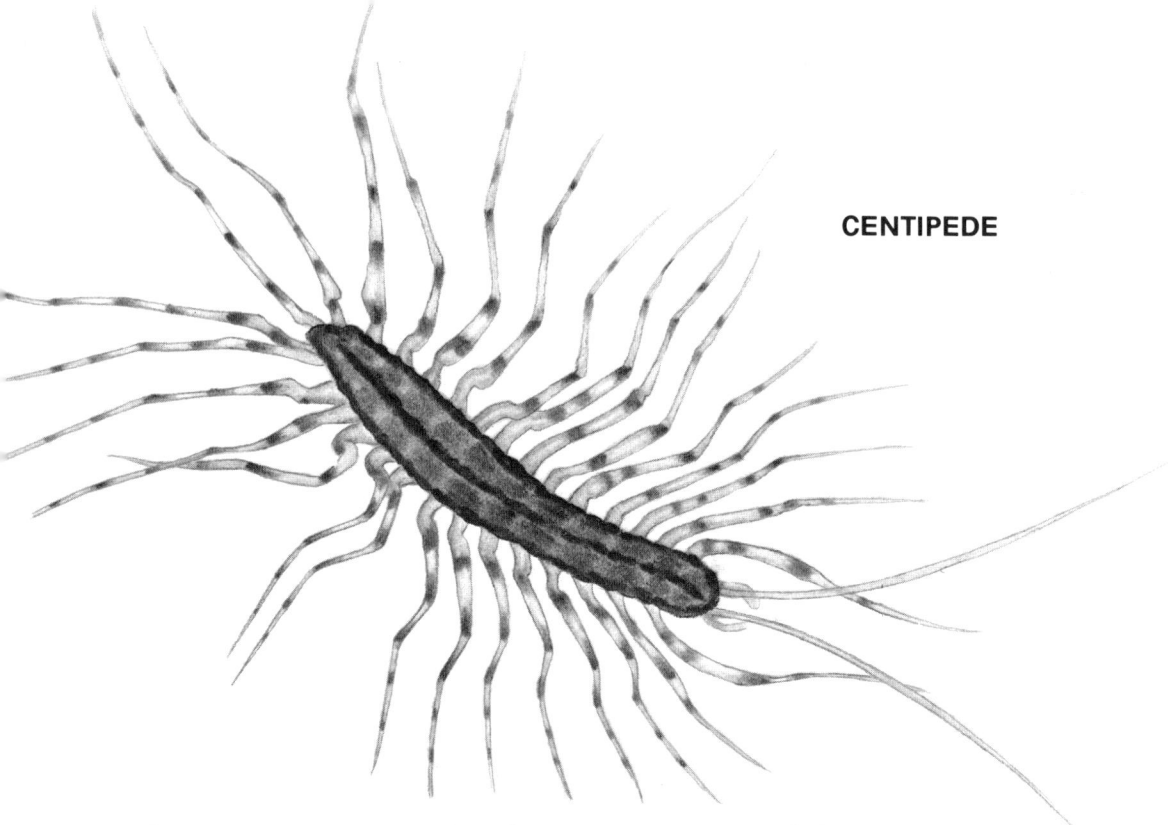

Other small animals that you may find in grassy spots or under a rock are centipedes, millipedes, earthworms, and pill bugs.

Centipedes and millipedes have many little legs. Centipedes will usually run away when you uncover them. Millipedes will usually curl into a ball.

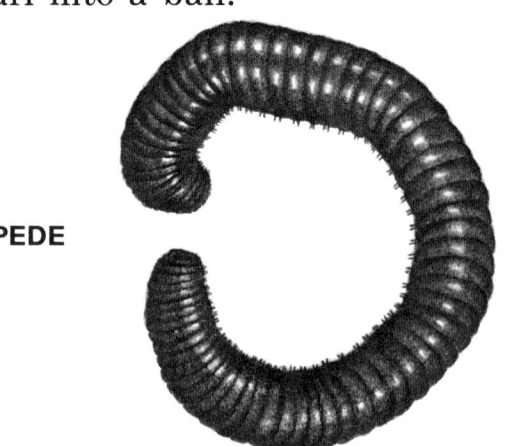

MILLIPEDE

Earthworms live in the soil.
In sunlight, their bodies dry up and they die.
But after a heavy rain, you will find many
earthworms on the ground.

Here is how you can discover some things that earthworms do.
Collect a few earthworms in a jar
along with some moist leaves or grasses.

EARTHWORM

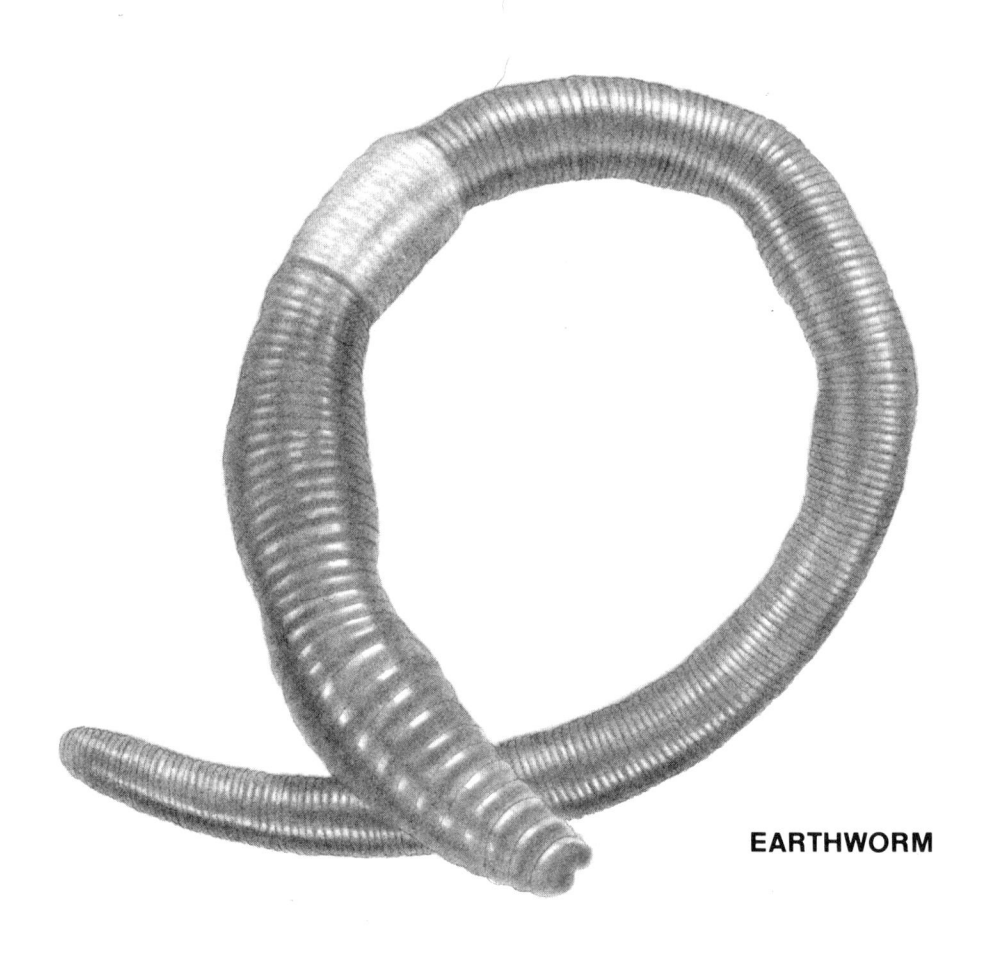

At home, wet a paper towel.
Take an earthworm out of the jar
and place it on the towel.
Observe what it does.
How does the earthworm move?
Run your fingers along the underside of its body.
Do you feel tiny bristles?
How do you think they help the earthworm move?

Now try this.
Turn off all the lights in your room.
Shine a flashlight on one side of the earthworm.
What happens?
How would moving away from a strong light help an earthworm to live longer?

Earthworms burrow through the soil eating any bits of animal or plant materials they come across.
In this way, they loosen up the soil
and make it easier for plants to grow.

EARTHWORM IN SOIL

PILL BUG CURLED UP

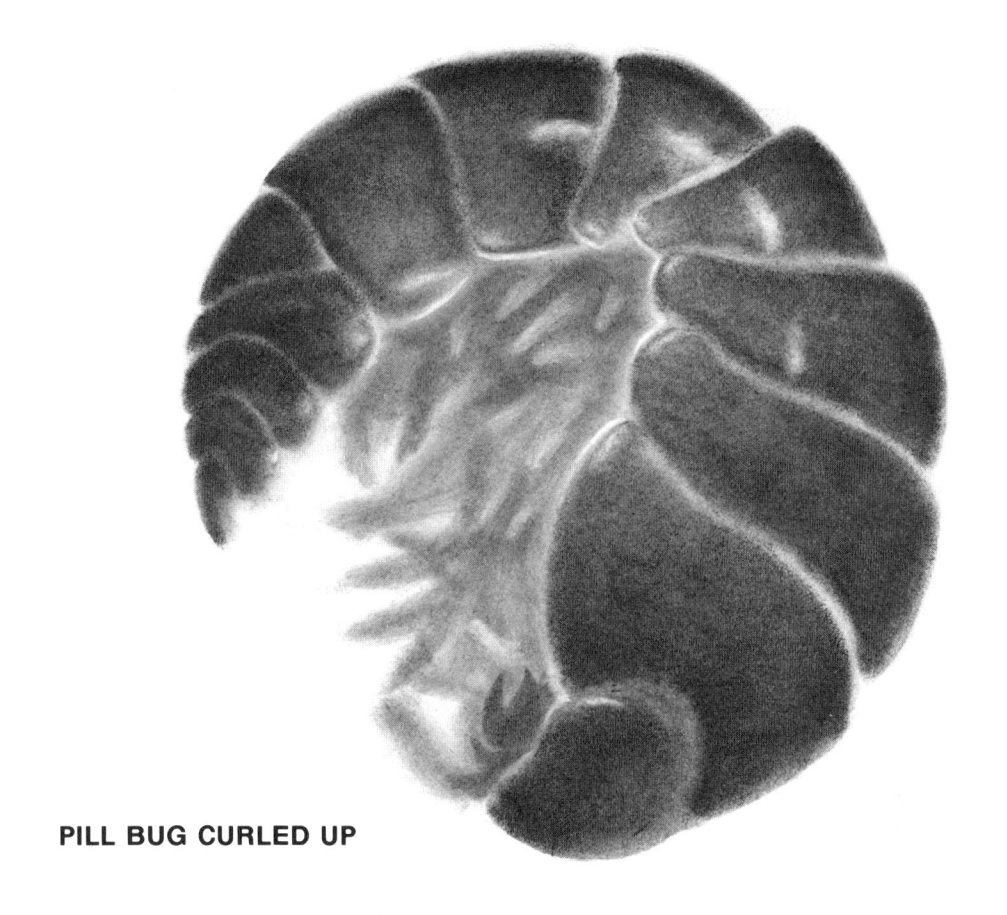

You can find pill bugs in damp places in a basement
or in a decaying log.
As soon as you uncover them, they will roll up
into a tiny ball or pill.
They are harmless animals that feed on
decaying plant and animal materials.

Like insects, birds are hatched from eggs.
Like you, birds have two legs.
But birds are different from all other animals
in a special way.
Birds are the only animals that have feathers.

HOUSE SPARROW

BLACK-CAPPED CHICKADEE

HUMMINGBIRD

There are many different kinds of birds.
Some birds, such as turkeys and chickens,
are too heavy to fly much.
Other birds, such as some kinds of hummingbirds,
weigh about as much as a nickel, and fly around
most of the day.

Some birds live in and around your neighborhood
all year round.
Others come just during the warmer weather.
Spring is a good time to look for birds.
In the spring, many birds
fly north to their summer nesting places.
They often fly at night and eat and rest during the day.

SCARLET TANAGER

STARLING

Early morning is a good time to look for birds.
Some birds walk.
Others hop.
Follow bird tracks in wet mud.
How is a walking track different from a hopping track?

Birds spend much of their time eating.
You may see them looking for food
on lawns, in trees, or in vacant lots.
Watch for scarlet tanagers, warblers, or nuthatches
searching for insects in trees.
Watch for robins gathering earthworms on lawns.
Watch for pigeons and starlings eating
almost anything they find.

NUTHATCHES

BLUE JAY

Some of the insects that birds eat can harm the leaves and branches of trees.
By eating insects, birds help to protect trees.
Trees help birds, too.
Many birds build nests high in the branches of a tree.
A high place is a safer spot for eggs and young birds.

During a snowy winter, insects and seeds are scarce.
Birds that stay in your neighborhood
must search all over for food.

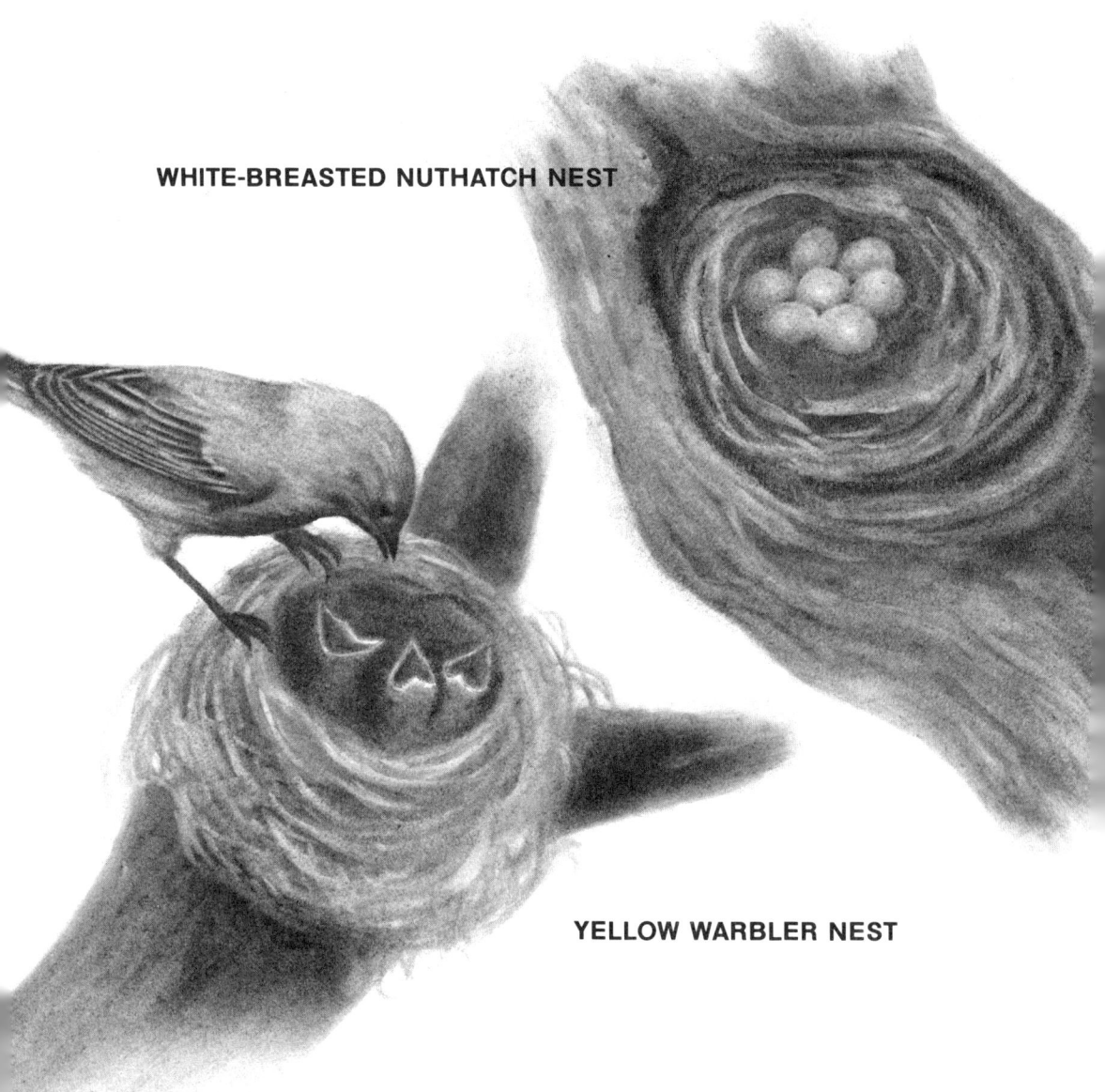

WHITE-BREASTED NUTHATCH NEST

YELLOW WARBLER NEST

Look for nesting birds in trees, on windowsills, fire escapes,
and under the overhangs of buildings.
You also may see bird nests that have been deserted.

PIGEON NEST   HOUSE SPARROW NEST

A robin's nest is made of twigs and mud.
The upper part is lined with leaves, string, and even rags.
Sparrows make a nest of grass and feathers,
usually under the overhang of a building.
Doves may nest on just a few twigs or sticks on a flat place.
What other kinds of nests can you find?
Do not disturb a nest even if it is deserted.
Some birds will use the same nest year after year.

ROBIN NEST

If you set up a feeder, you can help birds
during cold weather.
Place the feeder in a protected spot,
so that cats and squirrels cannot get to it.
Most wintering birds will eat seeds or suet.
Watch the birds as they come to the feeder.
Which foods do they eat first?

Do different kinds of birds eat the same foods?
What other things do you observe about the way birds eat?

WINDOW FEEDER

HOUSE SPARROW

Look at the shape of a bird's beak.
It will often give you a clue as to the kind of food it eats.
A sparrow's wedge-shaped beak is just right
for cracking seeds.
A robin's long beak helps it catch earthworms in the soil.

ROBIN

What kinds of sounds do birds make?
Can you tell the difference between their
songs and their calling sounds?

Listen to the robin.
Its song sounds like lee-a-lee
but its call is more of a harsh chirp.
Birds sing to warn other birds away
from their territory.
They call to each other to keep in touch
and warn of danger.

SPARROWS   **STARLING**

Listen to the starling and the pigeon.
Do they sing?
Do they make any other sounds?
What about the common sparrow?
See if you can learn its songs.
When do you hear most bird sounds?
When you hear a bird, see if you can locate it.
Notice its size, its shape, and its colors.
Then look it up in a bird book.
You may be surprised at the number of birds that you find in your neighborhood.

PIGEONS

GRAY SQUIRREL

Are there squirrels in your neighborhood?
Where do you think squirrels live?
Look up in the branches of tall trees.
Look on the ground around the trunks of trees.
Squirrels are good tree climbers.
See how they use their sharp claws to cling to the branches.
How does a squirrel get from tree to tree?
Watch a squirrel jump.
Its springy hind legs are just right for running and jumping.
It uses its tail to balance itself as it leaps.

Perhaps you can put out a peanut for a squirrel to eat.
Many park squirrels will come over and beg for food.
Do not try to touch a squirrel or feed it with your hands.
Squirrels have sharp teeth and may bite
if they are frightened.
How does a squirrel pick up a nut?
What does it do with it?
What other kinds of foods do you see squirrels eating?
In the fall, a squirrel may bury a peanut
along with the acorns and seeds it gathers.
The squirrel will use the stored seeds and nuts
as food during the winter.

As winter approaches, a squirrel begins to eat more and more.
Its body becomes plump and its fur begins to thicken.
Layers of fat beneath its skin and bushy fur
help a squirrel stay warm during the cold days of winter.
Did you ever watch a squirrel fluff out its tail?
It presses the fluffy tail against its back
to help it keep warm.
In the spring, a squirrel looks shabby.
Its thick fur coat is beginning to shed.
It is losing the extra hair that helped it keep warm
during the winter.
There are few nuts and seeds to be found during the spring.
Now squirrels are on the lookout for the tender shoots
of new branches and young trees.

**GRAY SQUIRREL
FLUFFED OUT**

Of course, most squirrels are always ready for a handout
of some peanuts or a crust of bread.

GRAY SQUIRREL NEST

If you look high in the branches of a tree
you may be able to see a squirrel's nest.
It looks like a clump of leaves and twigs.
This kind of nest is used in warm weather or
where there is a short supply of holes in trees.
A leaf and twig nest is not very sturdy.
A hard rainstorm will often destroy such a nest.
Most squirrels find shelter in holes in trees
or even under the overhang of a roof
rather than in a nest of leaves.
The young are born and spend the first few weeks
of their lives in the tree holes.

GRAY SQUIRREL

A squirrel has to live near a tree.
It provides a squirrel with food and shelter.
It is also a good place to flee
from a cat or a dog or some other danger.
You will not often find a squirrel more than
a few feet from the nearest tree.
Squirrels and trees go together.

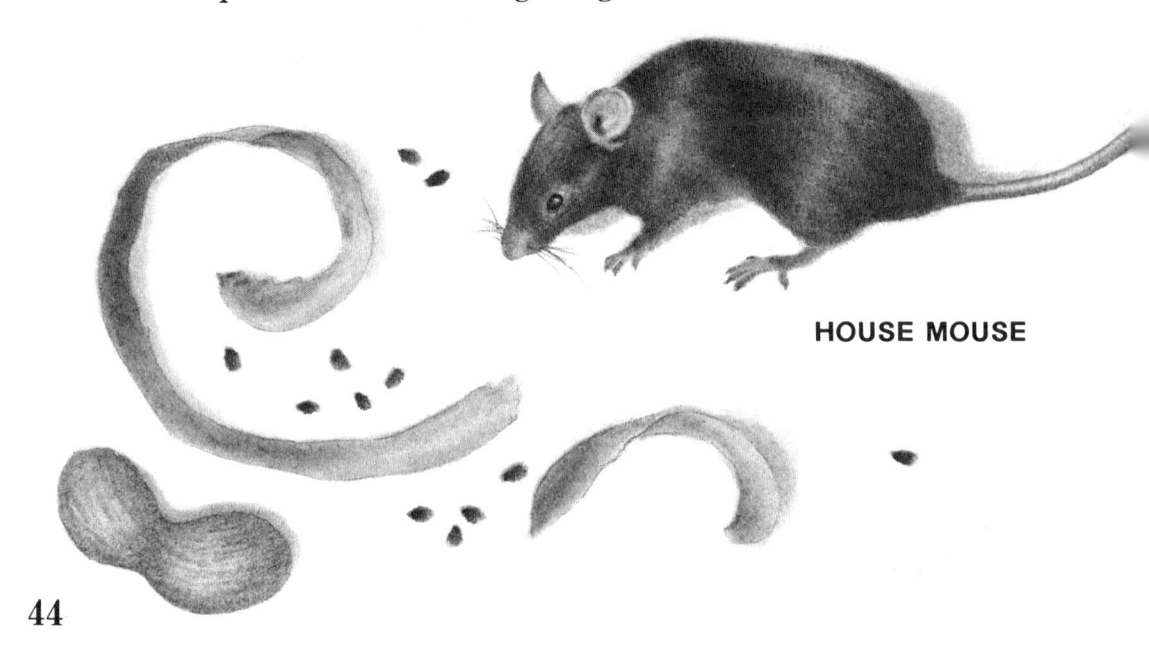

HOUSE MOUSE

Rats and mice are close relatives of squirrels.
But you may not spot them as easily as squirrels.
Most of the time, rats and mice come out at night.
During the day they stay in narrow cracks and holes
in the ground, in trees, and in the walls of houses.
Rats and mice eat almost anything.
Because they carry germs and sometimes bite,
most people consider them pests and try to get rid of them.

RATS

Some animals in your neighborhood are pets.
They are cared for by people.
Other animals are wild.
They care for themselves.
Make a list of all the animals that you find
in your neighborhood.
What are they doing when you see them?
Do you see them in the morning or in the evening?
Which are all year residents and which are visitors?

Animals live in neighborhoods just as you do.
But animal neighborhoods are called *habitats*.
A habitat is a place where an animal lives and where it
can find the food, shelter, and other things it needs.
When you walk in your neighborhood,
you are also exploring animal habitats.
Next time you take a walk, see if you can tell
the animal habitats that make up your neighborhood.